CHEMOPHOBIA CLUSTER

at the

Natural Resources Defense Council

AF731037

By **Susan Z. Ingber** and **Gilbert Ross, M.D.**

AMERICAN COUNCIL
ON SCIENCE AND HEALTH

September 2011

ACSH accepts unrestricted grants on the condition that it is solely responsible for the conduct of its research and the dissemination of its work to the public. The organization does not perform proprietary research, nor does it accept support from individual corporations for specific research projects. All contributions to ACSH—a publicly funded organization under Section 501(c)(3) of the Internal Revenue Code—are tax deductible.

Individual copies of this report are available at a cost of $5.00. Reduced prices for 10 or more copies are available upon request.

Published by the American Council on Science and Health, Inc.—September 2011.

Contents

Introduction to the Scare

In March 2011, the Natural Resources Defense Council (NRDC), an environmental activist group, released a report alleging that chemical exposures throughout the country have led to numerous "disease clusters." NRDC Senior Scientist Gina Solomon, MD, MPH, testified before the United States Senate Committee on Environment and Public Works on March 29th in order to promote the need for far-reaching changes in the Toxic Substances Control Act (TSCA) of 1976 by referring to this report as the basis for NRDC's call for TSCA "reform."

TSCA was enacted to protect the public from exposure to potentially harmful chemicals in the environment. The NRDC's version of reform would include allocating massively increased funding for federal agencies — especially the Environmental Protection Agency (EPA) — to investigate and deal with other possible disease clusters; reform would also include more stringent standards for chemical emissions and releases into the environment, based on the **precautionary principle**. This principle states, simply, that if there is any concern that a substance may cause human or environmental harm — no matter the evidence for concern, and no matter the level of exposure — then the substance should be removed from commerce until its safety can be proven. Such "reform" would place huge financial burdens on both chemical manufacturers and American taxpayers, given the thousands of chemicals on the market and the extent of "concern" generated by agenda-driven pseudo-scientific warnings such as this one.

In response, the American Council on Science and Health (ACSH) decided to use the principles of sound science — specifically, the Bradford Hill criteria for causation (1965)[1] and Guzelian *et al.*, "Evidence-based toxicology" (2005)— to critically evaluate the NRDC's purported disease clusters to determine if there is any evidence-based support for these claims and, if so, to what extent they relate to TSCA reform.

What's a Disease Cluster?

Disease clusters are a frequent topic of media attention, and have come to have two different meanings — one for the public and one for scientists. While the public typically thinks of disease clusters in terms of disease caused by industrial pollution, scientists tend to see the issue differently. Some epidemiologists define a disease cluster as a geographic area, time period, or group of people with a greater than expected number of cases of a given disease.

Epidemiologists — scientists who study the causes and distribution of human diseases — expect disease rates to vary slightly from year to year, and use statistical tests to determine whether a given rate is different enough from the average to qualify as unexpected.

Proving that a cluster exists entails a rigorous statistical analysis, in which the number of cases observed in some area over a specified interval of time is compared to the expected number of cases for that time period and area. Sci-

1 See Appendix I for the full Bradford Hill criteria.

entists can use clustering as a way to learn about associations between disease and any external agent, not only pollutants.

The public and the media, on the other hand, often use the term "disease cluster" in a more general way. When members of a community believe that the number of disease cases in their community is abnormally high, they will often seek an explanation for the "disease cluster." The media may then report on efforts to explain the cluster, before a statistical analysis has verified that the rates of disease in the community are actually elevated.

Thus, the term "disease cluster" is often used by the public and the media to refer to the perception of an elevated number of disease cases, and by epidemiologists to refer to confirmation of this perception. The community perception may reflect an elevated rate of disease, or it may not.

Methods

Our report investigates the scientific bases — or lack thereof — among the 42 clusters in 13 states that NRDC linked to one or more chemical exposures (or, in the case of one alleged cluster, electromagnetic fields) alleged to have occurred since TSCA. Clusters for which federal, state, or local government investigations have not yet been completed are identified and described separately. In our report, clusters from the same county were consolidated into a single listing.

Results

With few exceptions, the NRDC's findings fly in the face of the conclusions reached by objective governmental public health agencies, such as the Agency for Toxic Substances and Disease Registry (ATSDR), an affiliate of the U.S. Department of Health and Human Services, and the U.S. Centers for Disease Control and Prevention (CDC), along with various state departments of public health.

Valid clusters

Among the 42 clusters cited by NRDC, only one case appeared to be based on valid epidemiological principles (e.g. a chemical known to be linked to a particular condition or disease). This case involved asbestos exposures in Libby, Montana. Vermiculite asbestos was mined there from the 1920s to 1990, leading to asbestos contamination in the households of miners in addition to excessive occupational exposures. Note that, in addition to the markedly elevated levels of the rare cancer mesothelioma (known to be associated with excessive exposure to asbestos), ATSDR found a 40- to 60-fold increased risk of death from asbestosis in Libby area residents. The first asbestos regulations were, in fact, put into place in 1973 under the Clean Air Act after the EPA identified asbestos as a significantly hazardous contaminant in 1971.

Possible clusters

ACSH found that four of the clusters listed in NRDC's report might be reasonably linked to environmental exposures. All of them, however, require follow-up with more extensive observational studies that directly measure residents' exposures to the substances in question in order to gather more evidence to either confirm or reject a link between the alleged chemical exposure and the associated diseases.

Ongoing investigations

NRDC included 13 alleged clusters whose official public health investigations are still in progress. ACSH chose not to assess the legitimacy of these claims without data from completed governmental evaluations.

Purported clusters lacking supportive evidence

The NRDC compilation lists several alleged clusters in which the investigations undertaken by public health authorities were finally inconclusive and abandoned. If a comprehensive epidemiological assessment, such as those conducted by the ATSDR, CDC, or state health departments, cannot justify further investigation into an environmental chemical exposure, then the NRDC's assertions that a disease cluster is linked to chemicals remains unsupported.

NRDC's report also includes "clusters" that state public health departments actually ruled out. Moreover, despite the NRDC's claim that they listed only disease clusters that occurred after the implementation of TSCA, their report actually includes several instances in which exposures occurred before 1976. Our investigation into the remaining 24 alleged disease clusters revealed them to be entirely without merit.

Conclusion

The large majority of the claims in the NRDC's report have no scientific basis and thus fail to support NRDC's assertion that TSCA has not provided adequate protection from potentially harmful chemical exposures.

NRDC-Proposed Disease Clusters By State

*For a list of clusters by category, see Appendix II

ARKANSAS

1. Prairie Grove, AR

NRDC's claim: A nearby poultry feed manufacturer and poultry farm exposed the community to dangerous levels of arsenic, a component of their chicken feed leading to a cluster of testicular cancer. NRDC also blamed a closed-down nuclear reactor and low-level radioactive landfill as possible cancer threats.

Origin of the claim: In 2001, the Arkansas Department of Health identified a cluster of five testicular cancers between 1997 and 2001; three of the cases occurred in 14-year-old boys. Local residents were concerned that the poultry factories were contributing to the town's "high rates of cancer and other health problems" because arsenic-contaminated chicken manure was used as fertilizer and spread on fields near schools and homes.

The science: Dr. David Bourne of the Arkansas Department of Health stated that the number of testicular cancer cases diagnosed within five years was "higher than expected and therefore troublesome." However, there is limited research into the potential link between arsenic and testicular cancer. "Almost nothing is known regarding health effects of organic arsenic compounds [used in the chicken feed] in humans," according to ATSDR. The ATSDR investigation did not link the testicular cancers to chemicals — nor did they rule it out.

ACSH assessment: The cause of the testicular cancer cluster has not been determined.

CALIFORNIA

2. Earlimart, McFarland, and Rosamond, CA

NRDC's claim: Pesticide use in these farm regions led to a cluster of various childhood cancers, including brain, kidney, and muscle cancer.

Origin of the claim: In response to pressure from the United Farmworkers (UWF) union, the California Department of Health Services (CDHS) launched an investigation into childhood cancer incidence in the Fresno, Kings, Kern and Tulare counties of California. UFW noted that all of the cases occurred in children of farm worker families. In the Tulare county town of Earlimart, there were five childhood cancers among the 1,050 resident children between 1986 and 1989. Between the years 1975-1984, eight cases of childhood cancer (ages 0-15) occurred in Rosamond (Kern County); half of them were due to a rare form of brain cancer called medulloblastoma. Meanwhile, neighboring McFarland (Kern County) experienced a cluster of 13 childhood cancer cases within a similar time period.

***The science*:** According to CDHS, the rate at which children in Rosamond developed cancer was several times greater than in areas like Los Angeles and San Francisco. The agency could not, however, link an environmental cause to any of the childhood cancer clusters. While the NRDC claims that farms' under-reporting of "restricted pesticide use" prevented the CDHS from linking the cancers to chemical exposures, CDHS was unable to detect an excess of carcinogens in the air, soil, or water that could account for the clusters. Though CDHS did detect dioxins, furans, and other potential carcinogens in Rosamond, for example, they could not determine any plausible ways for children to have been in contact with these chemicals.

***ACSH assessment*:** There is no scientific evidence that can link the childhood cancer clusters in Earlimart, McFarland, and Rosamond, CA to pesticides or other environmental chemical exposures.

3. Hinkley, CA

***NRDC's claim*:** Pacific Gas & Electric (PG&E) leached hexavalent chromium into the town's water supply, leading to a cluster of various diseases (breast cancer, Hodgkin's disease, miscarriages, and spinal deterioration).

***Origin of the claim*:** Made famous by the film *Erin Brockovich* (starring Julia Roberts), this case involves a small town that brought a lawsuit against PG&E for allowing hexavalent chromium to leach from their plant reservoirs into residential drinking water between 1952 and 1966. The residents blamed a variety of major health effects on this alleged contamination, even though many of them manifested decades later. Erin Brockovich and community residents won a $330 million settlement in 1996.

***The science*:** Despite the much-publicized litigation and media attention, a series of three studies completed by the California Cancer Registry found no increase in cancer occurrence from 1988 to 2008. Meanwhile, the NRDC tries to discredit these conclusions by stating that some state officials believe that the population size is too small for cancer surveys to yield conclusive results, describing the Hinkley case as "an example of why disease clusters are difficult to prove." The capricious, unscientific basis of this claim is evidenced by the observed count of new cancer cases in the Hinkley area population that was, in fact, lower than expected for the demographic characteristics of the population.

There is an abundance of peer-reviewed research supporting the hypothesis that hexavalent chromium did not cause the numerous ailments claimed in the Hinkley lawsuit. For example, hexavalent chromium is not readily soluble in water (only 34 percent of the chromium found in water is in the hexavalent form, most of which settles to the bottom of the water supply, according to the EPA), and what little that does dissolve is converted to a harmless form called trivalent chromium. Trivalent chromium is required for glucose metabolism and is therefore essential for life; excess levels are quickly excreted.

Finally, this contamination occurred prior to TSCA, and is thus irrelevant to the call for "reform."

***ACSH assessment*:** Hexavalent chromium did not cause the diseases found in Hinkley, nor was there evidence of elevated cancer incidence during the time in question.

4. Kettleman City, CA

***NRDC's claim*:** A nearby hazardous waste disposal facility caused a cluster of birth defects (of various types) observed between 2007 and 2010.

***Origin of the claim*:** The California EPA (CalEPA) investigated a cluster of 11 children born with major structural birth defects such as cleft palate, heart and brain problems, and limb defects, to mothers who resided in Kettleman City during their pregnancies. The residents believed that exposure to pollutants from a hazardous waste disposal facility located less than four miles away was to blame for the malformations.

***The science*:** Though CalEPA was able to detect pollutants in the city's air, water, and soil, it could not link the birth defects to any particular environmental exposure. The California Department of Public Health (CDPH) concluded that all of the birth defects represented different underlying conditions, though a small number coincidentally shared some features.

ACSH assessment: The reason or reasons for the elevated incidence of birth defects in Kettleman City is undetermined, but is not related to the waste disposal facility.

5. Montecito, CA

NRDC's claim: A childhood leukemia and lymphoma cluster is linked to electromagnetic fields (EMFs) from a transformer station near the Montecito Union School.

Origin of the claim: CDHS confirmed a cluster of childhood leukemia and lymphoma in Montecito from 1981 to 1988 that was found to be at a rate that, according to NRDC, is "five times higher than would be expected during an eight-year period in a city of its size." The community expressed concern that EMFs produced by a transformer station next to the school and by a nearby transmission line that crossed over the school property were to blame.

The science: A survey by independent consulting firm Enertech Inc., chosen by community parents and overseen by the CDHS found that the magnetic field levels at the school were not unusually high. The researchers concluded that the exposure levels, which ranged from five to 30 milligauss near power lines, were "similar to what one is exposed to near a common electrical household appliance such as a TV or radio." CDHS has not been able to determine particular cause for the cluster.

ACSH assessment: There is no evidence to support the alleged link between the EMFs and the childhood leukemias and lymphomas. Moreover, the cause of this supposed cluster is not even a chemical exposure and therefore cannot be relevant to TSCA — yet NRDC still had no compunction about using it to support their agenda.

6. Oroville, CA

NRDC's claim: A pancreatic cancer cluster occurred as a result of well water contamination by Koppers wood treatment facility.

***Origin of the claim*:** The CDPH declared that the 24 cases of pancreatic cancer found in Oroville in 2004 and 2005 amounted to twice the expected rate of six cases per year. Residents feared that a fire at the Koppers wood treatment facility lead to well water contamination with pentachlorophenol and related chemicals.

***The science*:** The CDPH could not link the increased incidence of pancreatic cancer to well water contamination by Koppers wood treatment facility: "Although a variety of possible environmental and occupational exposures were noted, none were consistently found among enough [pancreatic cancer patients] to explain the occurrence of the excess." The agency also found a number of confounding factors that could have affected the elevated pancreatic cancer rate. "Only one case had ever worked at the wood treatment facility," according to the CDPH. "[And] only two lived in an area evacuated during the fire; and the only person who may have consumed well water from the groundwater plume was the former worker... Many cases shared a known risk factor for pancreatic cancer (such as smoking, being overweight, having diabetes, or a family history of pancreatic cancer), and some worked at specific occupations that may be linked to pancreatic cancer (such as mechanic, welding, working with pesticides)."

***ACSH assessment*:** The pancreatic cancer cluster — if indeed there was such — is unrelated to Oroville's well water.

7. Santa Susana, CA

***NRDC's claim*:** Water and soil contaminated by the Santa Susana Field Laboratory (SSFL) caused thyroid and bladder cancer clusters.

***Origin of the claim*:** A University of Michigan study linked thyroid cancer risk to residents' distance from SSFL, which NRDC calls "a notorious source of widespread radioactive and chemical contamination." They blame perchlorate, based on its ability to cause thyroid tumors in laboratory animals and its detection in groundwater in areas surrounding SSFL.

***The science*:** The Michigan researchers acknowledge a number of limitations in their study:

> It is important to recognize that associations observed between distance from SSFL and the incidence of specific cancers are based on small numbers of cases in the region closest to SSFL. Thus, these associations are estimated imprecisely and may represent chance findings. In addition, observed associations may have been biased by certain methodologic limitations — use of distance from SSFL as a crude proxy measure for environmental exposures, mobility of the residential population before and during the follow-up period, and lack of information [about] other cancer risk factors...that might distort the observed associations.

ATSDR's preliminary research into the likelihood that residents' exposure to SSFL contaminants can be linked to the elevated bladder and thyroid cancers indicate that "it is unlikely that people living in communities near the site have been exposed to substances from the site at levels that would have resulted in adverse health effects... Changes in site operations, such as reduced frequency of rocket engine testing, discontinuation of trichloroethylene use, and shut down of nuclear operations, make it unlikely that future exposures to the offsite community will occur."

***ACSH assessment*:** While ATSDR has recommended that updated demographic and cancer data be evaluated to determine if thyroid and bladder cancer elevations persist, the data indicate that residential exposure to SSFL contaminants is most unlikely to have caused the cancer clusters.

DELAWARE

8. Millsboro, DE

***NRDC's claim*:** Lung cancer cluster from contamination of the drinking water supply by the nearby Indian River Power Plant's coal ash landfill. Contaminants include arsenic, chromium, and thallium.

***Origin of claim*:** Out of concern over water contamination from the Indian Power Plant, Millsboro residents requested that the state of Delaware launch an investigation into whether the community was the site of a cancer cluster.

***The science*:** The Delaware Department of Public Health (DDPH) found "an incidence of 553.99 cancer cases per 100,000 residents of this area between 2000 and 2004 compared with the Delaware state rate of 501.3 and the U.S. rate of 473.6 cancer cases per 100,000 residents. Thus this study confirmed that the rate of cancer cases in this area is 17 percent higher than the national average." However, the DDPH ultimately concluded that it did *not* find a cancer cluster, noting that "only lung cancer was elevated for that area ... well water testing did not indicate a link between well water contamination and cancer rates for the area."

A separate "Indian River area" retrospective study evaluating all-cancer incidence in Millsboro and surrounding areas was also carried out. It found that both elevated lung and bladder cancer cases among males in the area accounted for the overall increased cancer risk compared to the state average. The authors made no associations with environmental factors and did not conclude that there were any "cancer clusters." The overall lung cancer incidence in the area was 19.5 percent (compared to 15 percent in the state of Delaware). The researchers hypothesized that tobacco use and migration from other states could have contributed to the elevated lung cancer rate. "However, evidence that coal burning power plants specifically cause cancer is not clear... It is not known if the prevailing winds would deposit respirable particulates in the Indian River community or if such particles would be carried east out to the ocean..."

ACSH assessment: The DDPH declared that the elevated lung cancers do not, in fact, constitute a cancer cluster, and they believe that residents' exposure to the byproducts of coal burning power plants is highly unlikely to be of sufficient intensity to pose an increased health threat; further, other more likely health risks, such as smoking, were not accounted for.

FLORIDA

9. Immokalee, FL (and North Carolina)

NRDC's claim: Birth defects associated with pesticide exposure from agricultural work on Ag-Mart Produce, Inc. tomato fields here and in North Carolina by three families.

Origin of claim: In February 2005, the Healthy Start program in the Collier County Health Department (CCHD) in Florida identified three infants with congenital anomalies who were born within 8 weeks of one another and determined that all three mothers had worked for the same tomato grower — Ag-Mart Produce, Inc. All three women had worked on the grower's Florida farms in 2004 before transferring to its North Carolina farms later that year.

The science: Florida and North Carolina health officials found only a "plausible association" between the pesticide exposures of three women to birth defects in their respective babies. However, these cases could (possibly) be the result of safety violations by the Ag-Mart Produce, Inc. (The North Carolina Department of Agriculture and Consumer Services alleged that Ag-Mart Produce, Inc. had 369 pesticide violations, "including the use of six pesticides classified by the [EPA] as among the most dangerous to workers and applying a dangerous pesticide three times more often than allowed by law," according to the NRDC's reporting.) In one case in particular, the mother, whose child was born with limb defects, may have worked as many as 256 hours during times of "restricted entry," during which a number of pesticides known to cause limb defects in animals were used. The other case mothers' pregnancy outcomes either failed to coincide with exposures to teratogenic pesticides or were more likely related to genetic predisposition. The North Carolina DPH concluded:

> While it is possible that the birth defects are unrelated to the case-mothers' occupational exposures, there is evidence, based on interview information and regulatory compliance information from the NC Departments of Agriculture and Labor, that the women's work environment likely put them at an increased risk of over-exposure to pesticides. In conclusion, the findings of this investigation warrant

concern and action on the part of public health and regulatory agencies charged with protecting the health of farmworkers.

ACSH assessment: Poor enforcement of established safety protocols may have put one child at increased risk for birth defects, but the evidence does not support a link to pesticide exposure for the other two cases. Enforcement of TSCA would have sufficed to prevent this single episode.

10. Loxahatchee, FL

NRDC's claim: A pediatric brain cancer cluster resulted from chemical leaks and spills by the rocket and jet engine company Pratt & Whitney.

Origin of the claim: A community group found that, since 1996, eighteen children had been diagnosed with brain cancer, and three with brain cysts. Residents blamed Pratt & Whitney for "leaks and spills of chemicals, such as solvents and pesticides, on its 7,000 acres for the last 30 years."

The science: The CDC reviewed a number of studies, including a 2009 Florida Department of Health (FLDOH) cancer investigation (the "Acreage cancer review") and case control data analysis, along with 2010 water, soil, and radon samplings by the FL Department of Environmental Protection. In sum, the CDC concluded:

> Based on the data provided, we concur with the conclusion of the FLDOH that the investigation (including the case control study and environmental testing) did not reveal any potential risk factors for which exposure among cases was statistically significantly higher than exposure among the controls. We do not have any recommendations for additional epidemiologic studies or environmental testing at this time. We hope that the extensive environmental testing and the case-control study have provided the residents of the Acreage with valuable scientific information to help allay their concerns about environmental contaminants in their community.

ACSH assessment: There is no evidence to support any association between the cluster of pediatric brain cancers and exposure to certain environmental chemicals.

11. Tallevast, FL

***NRDC's claim*:** Various cancers linked to long-term use of "contaminated" groundwater by the American Beryllium Company (ABC) between 1974 and 2006.

***Origin of the claim*:** In 2008, ATSDR determined that previous long-term use of Tallevast, FL groundwater for consumption and household purposes posed a public health risk. NRDC blames ABC, a manufacturer of machine parts from 1962 to 1996, because they allowed trichloroethylene (TCE) and "cancer-causing solvents" to leach into the groundwater supply.

***The science*:** FLDOH collaborated with ATSDR to determine whether TCE could have caused the uptick in Tallevast cancers. TCE was deemed the most significant potential health threat because it was present at the highest concentration of all the contaminants and had the highest "theoretical increased cancer risk." Counter to NRDC claims that the report concluded the residents were at an increased risk for developing kidney cancer, liver cancer, leukemia, and lymphoma, the FLDOH report actually specifies that this risk is only theoretical and does not reflect the actual cancer incidence measured in Tallevast. The report further specifies that "TCE is most strongly linked to liver cancers based on toxicology data and occupational studies. Less strongly linked have been kidney cancers, non-Hodgkin's lymphoma, and leukemia in children." Meanwhile, of the 83 cancer cases found during the study period, no liver cancer or leukemia (including childhood leukemia) cases were documented. The majority of the cases deemed "cancers of interest," which amounted to 38 cases, were of the lung (11 total cases, six among blacks) and prostate (17 total cases, 11 among blacks). Lung cancer has not been linked to TCE exposure, and prostate cancer has only been weakly associated with TCE exposure. Moreover, despite community concern that the heaviest exposure and contamination occurred in the early years of the ABC plant's operation, which in theory would have led to an uptick of cancer beginning in the mid- to late-1970s through the early 1980s, the FLDOH review found that the majority of the cancers occurred in the early 1990s. Thus, the cancer incidence does not correlate with the levels of TCE contamination.

***ACSH assessment*:** While the FLDOH recommended that a community health study involving interviews and review of patient medical records be conducted to more accurately detect cancer incidence and rule out confounding factors such as migration in and out of Tallevast, the data collected thus far render a link to TCE highly unlikely.

LOUISIANA

12. Amelia, LA

***NRDC's claim*:** The local Marine Shale Processor (MSP) leached chemicals into the air, soil, and groundwater, leading to a cluster of neuroblastoma.

***Origin of the claim*:** Louisiana State University Medical School researchers reported an excess number of neuroblastoma cases (five children) in the St. Mary Parish of Louisiana between 1986 and 1987. Driven largely by public concern that MSP's operations may have caused the excess neuroblastoma cases, city government and state health officials petitioned for an ATDSR public health assessment of MSP.

***The science*:** While the NRDC claims that a 1994 health investigation conducted by ATSDR concluded "there was evidence to suggest that adverse health outcomes in the community could be related to environmental exposure," a close examination of the ATSDR report's conclusions finds that the agency believes that "neuroblastoma has not been consistently linked to any single environmental agent." According to both the LSU researchers and ATSDR, St. Mary and surrounding parishes have "higher than expected" reported rates of malformations prior to both MSP's initial operations and the 1986-87 cluster, and the excess cases observed in 1986 occurred when several other northern Louisiana parishes experienced similar increases, making it difficult to determine whether the cluster is site-specific. The agency also notes that contamination of drinking water by MSP was "below levels of health concern."

Even LSU's assessment renders a link between the MSP's operations and the neuroblastoma cluster highly suspect. The researchers noted that "the cancer is usually thought to originate before birth and…three of the [five] stricken children were born before Marine Shale began operating."

Further, a Delta Toxicology examination of the neuroblastoma research found that "there are some familial aggregations of the disease, suggesting the potential of inherited susceptibility."

***ACSH assessment*:** The data do not support an association between neuroblastoma and an environmental exposure in relation to MSP's operations.

13. Coteau, LA

***NRDC's claim*:** A small cluster of childhood leukemia is linked to an unknown environmental chemical exposure.

***Origin of the claim*:** State health officials confirmed a cluster of childhood leukemia in the tiny community of Coteau after four children were diagnosed with leukemia.

***The science*:** In 2000, the Louisiana Office of Public Health conducted a case-control study of 40 children diagnosed with leukemia between 1983 and 1997 in the four-parish area of Lafayette, Vermilion, St. Martin, and Iberia identify risk factors associated with childhood leukemia in the area. Due to the small size of the study, however, epidemiologists could not draw any clear conclusions regarding potential environmental causes.

***ACSH assessment*:** The number of leukemias was too small to be a cluster, and there is no environmental link to the cancers.

14. Mossville, Calcasieu Parish, LA

***NRDC's claim*:** Dioxins and other organic chemicals emissions from several nearby petrochemical plants led to elevations of various diseases from 1988 to 1997.

***Origin of the claim*:** A 1998 door-to-door health survey of 96 Mossville residents conducted by the late University of Texas Medical Branch at Galveston toxicologist Dr. Marvin Legator (the same toxicologist who consulted on the widely publicized Love Canal case in New York, in which residents feared that the former toxic waste site posed a health threat) linked a variety of self-reported health problems to exposure to dioxins and other emissions from local petrochemical plants. The small town of Mossville was also featured in CNN's notorious "Toxic Towns" segment, hosted by Dr. Sanjay Gupta last year.

The science: The ATSDR followed up with their own investigation — which focused on cancer incidence — and extended the study period through the year 2004 (a total of 17 years of cancer incidence data) to include more data and increase the accuracy of the findings. The agency concluded, "There is no clear pattern indicating that Calcasieu parish has any consistently higher than expected rates for most cancers. The exceptions are melanoma of the skin, which was consistently elevated in whites, and cancer of the lung, which was consistently elevated in women." Although dioxins have been linked in some studies to cancers such as lung cancer, lymphomas, soft tissue sarcomas, and multiple myelomas, the data are not consistent across the board, and, according to the ATSDR's 2006 health consultation report, *Follow-up Exposure Investigation, Calcasieu Estuary (a.k.a. Mossville),* "Cancer in Calcasieu Parish cannot be predicted from the blood dioxin levels in the Exposure Investigation participants." Although, as NRDC points out, the ATSDR investigation did not explore the non-cancer ailments revealed in the University of Texas survey, the science has not been able to link dioxin exposure to the conditions found in the health survey.

ACSH assessment: The evidence does not support a link between dioxin exposure and the health issues seen in Mossville.

15. New Orleans, LA

NRDC's claim: A cluster of breast cancers resulted from exposure to a variety of chemicals, particularly polycyclic aromatic compounds (PAHs), associated with the former Agriculture Street Landfill. Private and low-income housing was built over this landfill in 1976.

Origin of the claim: A 2003 state Department of Health and Hospitals study showed a "statistically significant 50 percent excess of breast cancer" for all women in the census tract that includes the Agriculture Street Landfill area over the 10-year period ending in 1997. The area was declared a Superfund site in 1993.

The science: The ATSDR concluded in 2009: "The majority of the residential area and the Press Park Community Center has been classified as no apparent public health hazard since the levels of contaminants in the soil are generally below levels that may cause health problems. There are scattered pockets of lead, arsenic, and PAHs in soil that need to be addressed to limit any possibility of exposure to levels of health concern."

In light of these findings, NRDC insinuates that PAHs in the soil could have led to the breast cancer cluster, noting that, "There is evidence that PAHs can increase the risk of developing breast cancer." However, our research found that these data comes from animal studies, while human studies have not provided valid evidence of any link between PAHs and breast cancer. Meanwhile, ATSDR found that PAHs can increase the risk of lung, skin, and bladder cancers — but only at heavy occupational exposures.

ACSH assessment: Based on the soil measurements of a number of potentially hazardous chemicals associated with the landfill, along with the body of scientific literature on these substances, it is highly unlikely that exposure to the former Agriculture Street Landfill soil can account for the increased incidence of breast cancer.

MICHIGAN

16. Midland and Saginaw, MI

NRDC's claim: The Dow Chemical Company Midland plant allowed dioxins to contaminate the soil surrounding the Saginaw and Tittabawassee River, which led to a cluster of breast cancer.

Origin of the claim NRDC bases its cancer cluster claim on a 2008 retrospective study published in the journal *Environmental Health*. This study used geographical information system (GIS) mapping to analyze the geographical distribution of the approximately 4,600 breast cancers that occurred among residents living near the Tittabawassee and Saginaw River between 1985 and 2002. The study authors reported that their findings "suggest that neighborhoods in close proximity to the river were associated with a high risk of breast cancer… A positive association between possible exposure to environmental contamination and breast cancer (at significance level $P \leq .05$) was found in 40% of ZIP codes (e.g., in female populations residing in ZIP codes 48734, 48880, 48640, 48603, 48618, 48732, 48657, and 48604)." Meanwhile, the Michigan Department of Environmental Quality (MDEQ) found "elevated" levels of "dioxin-like chemicals" (DLC) in Saginaw River sediments, riverbank soils, and fish, and the Michigan Department of Community Health reported that fish consumption in these regions is a "public health hazard."

ACSH assessment: Researchers could not determine the actual levels of exposure experienced by the residents, and therefore could not document a clear link between DLCs in the Saginaw and Tittabawassee River areas and the breast cancer cluster. However, it is possible that this cluster can be linked to an environmental chemical exposure. Therefore, ACSH is classifying this as a possible cancer cluster, pending further investigation by a governmental public health department.

17. White Lake, Muskegon County, MI

NRDC's claim: Occupational and residential exposure to heavy metals and volatile organic compounds (VOCs) from various industrial facilities caused a cancer cluster.

Origin of the claim: Residents allege that companies such as Hooker/Occidental Chemical, DuPont, and the Whitehall Leather tannery have previously contaminated White Lake with heavy metals and VOCs that have led to a cancer cluster in their town.

The science: Concerned residents are compiling information on the residential and occupational history of people with cancer in the White Lake area for review by the Muskegon County Health Department. However, no cluster has been found. In fact, just weeks after NRDC published their disease cluster report, Ken Mahoney, chair of the Muskegon County Commissioners and vice-chair of the White Lake Public Advisory Committee, denied the disease cluster claim in a press conference. Mr. Mahoney explained that the media coverage alleging White Lake to be the site of a cancer cluster was in fact instigated by the baseless claims made by the "National Disease Cluster Coalition" (NDCC), an activist group that collaborated with NRDC on their report. NDCC, according to Mahoney, exaggerated the findings of the still-ongoing White Lake Area Cancer Mapping Project, charged with analyzing demographic data associated with cancer incidence. "If we were in fact a confirmed cancer cluster," said Mahoney, "we would be leading the charge to get attention to it and to get it resolved, but that is not the case."

***ACSH assessment*:** ACSH is classifying this case as a cluster of fear only, since no increased incidence of cancer has been found and there are no data to suggest that heavy metal and VOC exposures have led to health problems in White Lake.

MISSOURI

18. Herculaneum, MO

***NRDC's* claim:** A cluster of Amylotrophic Lateral Sclerosis (ALS) was caused by lead contamination in the soil and air around a local lead smelter

***Origin of the claim*:** The Missouri Department of Health and Senior Services (MDHSS) identified three ALS cases within a three-mile radius of a lead smelter in Herculaneum, MO (part of Jefferson County) between 1998 and 2002. The expected number was less than one case of ALS.

***The science*:** The MDHSS investigation was not designed to study risk factors of ALS. Therefore, while some studies have suggested that lead exposure is a potential environmental risk factor for ALS, the study could not conclude whether ALS cases in Herculaneum could be linked to the lead smelter, since such a link lacks scientific evidence. Another study referenced in the MDHSS report found that self-reported occupational lead exposure was a more important risk factor for ALS than was residential or recreational exposure. However, the MDHSS study did not test this hypothesis, and the study sample was too small to draw valid conclusions. Moreover, the MDHSS report concludes that prevalence of ALS in Jefferson County is comparable to rates in Western European counties observed in recent years.

***ACSH assessment*:** The number of ALS cases is too small to yield statistically significant results; there is no detectable relationship between lead and the three cases of ALS.

MONTANA

19. Libby, Montana

***NRDC's claim*:** Decades of asbestos-contaminated vermiculite led to a cluster of mesothelioma and respiratory disease

Origin of the claim: In 2008, the ATSDR identified a cluster of malignant cancers and respiratory diseases from 1979 to 1998 in Libby, Montana. Vermiculite contaminated with tremolite asbestos was mined in and near Libby from the 1920s through 1990 and was found to have contaminated the entire community — both occupationally and in the households of miners — and surrounding areas.

***The science*:** Tremolite asbestos has been linked to the following: malignant mesothelioma (cancer of the lining of the lung and abdominal cavity), asbestosis (when asbestos fibers become trapped in lungs, leading to scar tissue buildup and decreased lung function), and other non-cancerous complications associated with the lining of the lungs. ATSDR concluded:

> For the 20-year period reviewed in this report, mortality in Libby resulting from asbestosis was approximately 40 to 60 times higher than expected. Mesothelioma mortality was also elevated over this period, but since statistics on this extremely rare cancer are not routinely collected, it was difficult to precisely evaluate the degree to which mesothelioma mortality was elevated. Even though these two underlying causes of death are elevated in the Libby study population, they still may be underestimated when evaluated solely through the use of death certificate data. These findings support the community's

and the media's perception that Libby residents have been adversely impacted by historical asbestos exposure.

***ACSH assessment*:** The data provide strong evidence for a link between this asbestos exposure and asbestosis; although difficult to quantitate, mesothelioma was most likely linked to the mining. Therefore, this does appear to be a disease cluster linked to an environmental exposure, albeit exposure to a natural substance rather than a synthetic industrial chemical contaminant.

NORTH CAROLINA

20. Bynum, Chatham County, NC

***NRDC's claim*:** The widespread use of untreated river water led to elevated cancer rates from 1980 to 1985.

***Origin of the claim*:** NRDC cites a 1990 study that found above-average cancer rates (about 2.5 times higher than expected) in the Bynum area from 1980 to 1985. The NRDC report links the higher cancer rates to residents' consumption of untreated Haw River water from 1947 to 1976 — prior to the implementation of TSCA — after a typhoid fever epidemic had occurred.

***The science*:** Chatham County Health Director Holly Coleman announced shortly after the NRDC report was released that the cancer rates dropped below the state cancer rate once treated drinking water was provided to the town. "Cancer rates in that area of the county may have been an issue in 1980-85 as a result of the water source problem in the mid-century, but we have no data to support a higher rate of cancer in the county or in Bynum at this time," she said in a statement.

***ACSH assessment*:** Providing residents with treated drinking water and advising them against drinking river water may have helped to reduce Bynum's overall cancer incidence to within normal range. Although the exposure in question occurred before TSCA was put into place, thus making this cluster not strictly relevant to TSCA reform, its consequences did continue after 1976. Thus these findings might conceivably be construed as an example of water contamination contributing to a cancer cluster — although no specific chemical (or chemicals) is cited as a possible causative factor.

OHIO

21. Marion, OH

***NRDC's claim*:** A leukemia cluster is linked to arsenic and lead in the River Valley High School (RVHS) grounds.

***Origin of the claim*:** The Ohio Department of Health (ODH) found a cluster of leukemia in 1999 (83 cases, including nine RVHS graduates, between 1963 and 2000) in Marion, OH. ODH was asked to investigate whether contamination of the school grounds contributed to the elevated leukemia rate seen in the school's graduates. RVHS was built between 1961 and 1962 on the site of a former Army depot where cleaning and repairing of vehicles and heavy machinery took place.

***The science*:** The four-year ODH investigation into 47 of the 83 leukemia patients able to participate could not determine a particular cause for the increased leukemia. The researchers made three conclusions: 1) The most common factor among the study participants was direct or secondhand exposure to tobacco smoke; 2) While six of the nine River Valley graduates had extensive contact with school grounds through sports or agricultural activities, it is not known whether those exposures included potential cancer-causing contaminants; 3) Further investigation into the leukemia cluster is unlikely to identify additional factors that caused the leukemia.

***ACSH assessment*:** There is no link between the Army depot operations and leukemia, and the cluster could not be linked to a particular environmental exposure. Moreover, the former Army depot operations predate TSCA's implementation by more than a decade, rendering this alleged cluster irrelevant to the NRDC's call for "improved" chemical regulation.

22. Marysville, OH

***NRDC's claim*:** A cluster of eight leukemia cases between 1992 and 2001.

***Origin of the claim*:** NRDC did not attempt to link this small cluster — allegedly more than three-times the expected rate for Marysville residents aged 24 and under — to any particular environmental factor.

***The science*:** A 2007 study suggested that the uptick in leukemia could be linked to population changes between 1992 and 2001. However, the eight cases could not be linked to a specific cause.

***ACSH assessment*:** Eight leukemia cases over the course of ten years do not constitute a cancer cluster. Furthermore, NRDC does not even attempt to link these cases to any specific environmental chemical exposures.

TENNESSEE

23. Dickson, TN

***NRDC's claim*:** An oral cleft birth defects cluster is linked to trichloroethylene (TCE) contamination of well water and public water supplies and to release of toluene into the air.

***Origin of claim*:** Data collected by the CDC showed that between 1997 and 2000, 18 Dickson County babies with oral cleft palates were born. NRDC and several of Dickson County residents blame the defects on a number of local industries that dumped TCE drums in the Dickson County landfill between 1968 and 1977. They claim the TCE leached into and contaminated a private well and a public drinking water supply. This occurred before the implementation of "stringent landfill regulations and guidelines" (Tennessee's "Regulations Governing Solid Waste Processing and Disposal" in 1971; Tennessee's Hazardous Waste Management Program in 1980). NRDC also references the EPA's Toxic Release Inventory in 1997, which stated that Quebecor Printing released 1.4 million pounds of toluene into the air in Dickson.

***The science*:** According to a news article cited by NRDC, based on statewide statistics from 1989 through 1996, one case for every 1000 births was "expected." The CDC report found that while the Dickson County rate of these defects appeared elevated (on average 1.6 per 1,000 between 1989 and 1996), the defect rate for the county during this time period varied substantially from 5.42 per 1,000 births in 1989 to zero in 1993, 1995, and 1996, which was "expected," given the generally low number of live births in the county.

The CDC investigation did not examine the link between toluene and TCE and the birth defects alleged by residents. In fact, known risk factors for oral cleft palates (i.e., failure to use prenatal vitamins before and during pregnancy, smoking, family history) were found in each case, though "no one factor examined in this investigation could account for the increased rates in the county."

Moreover, while the NRDC claims that TCE has been linked to birth defects, only one study specifically evaluating oral cleft birth defects and TCE exposure from drinking water (published in 1995) could be found on the National Institutes of Health's published journal article search engine, PubMed, as of the publication of this report. In fact, the latest evaluation of TCE toxicity by the National Academy of Sciences and EPA in 2010 did *not* note a link between TCE and oral cleft defects. In addition, a search for such a link showed only a possible link to occupational exposures. The only other data come from animal studies. Oral cleft defects are among the most common birth defects and — while generally of unknown origin — have been linked in many instances to genetic (heritable) factors.

ACSH assessment: There is insufficient evidence to support a link between exposure to toluene, TCE, or their combination and the Dickson County oral cleft defects.

TEXAS

24. El Paso, TX

***NRDC's claim*:** A cluster of multiple sclerosis (MS) cases is linked to emissions from the ASARCO smelter facility one mile from Mesita Elementary School.

***Origin of the claim*:** In 1996 ATSDR and the Texas Department of State Health Services (TDSHS) found a two-and-a-half-fold increased risk of developing MS among residents who had attended Mesita Elementary School, based on national prevalence data.

***The science*:** This investigation was unable to determine a specific cause for the MS cluster. In addition, while genetic factors have been indicated as a causative factor for MS, there is a paucity of valid evidence linking heavy metal exposure to the disease. TDSHS has recommended that a national multi-site case control study be conducted to determine if there's an association between heavy metal exposure and MS.

***ACSH assessment*:** The exact causes of MS are not well known, and the El Paso investigation could not find a link between the MS and the ASARCO smelter facility.

25. Houston, TX

***NRDC's claim*:** A childhood leukemia cluster is associated with proximity to the Houston Ship Channel, specifically in areas with the highest benzene and 1,3-butadiene levels in the air.

***Origin of the claim*:** NRDC bases its claim on a single, non-peer reviewed, preliminary study commissioned by TDSHS and conducted by a team of University of Texas Health Science Center researchers. It found that children who live within two miles of the Houston Ship Channel have a 56 percent greater chance of developing leukemia than children living ten miles away.

***The science*:** The study attempted to correlate an estimated exposure to benzene and 1,3-butadiene (BD) with outdated cancer rates derived from medical records collected between 1995 and 2003. However, the normal expected risk of childhood leukemia is already very low, which means that even a few cases of leukemia can dramatically skew the cancer's frequency. The researchers assumed that the concentration of benzene and BD in the air was directly related to the distance from the Houston Ship Channel and state air pollution monitors, an assumption that fails to consider how wind and other weather patterns affect the chemicals' distribution in the air. Because the researchers were unable to perform what is called a randomized controlled clinical trial (the gold standard method of finding a cause-and-effect relationship in which two highly similar groups are compared after some intervention is given to one group but not the other), and since they could not measure individual test subjects' exposure to benzene and BD (e.g. with a blood test), this study falls far short of proving that these exposures pose a childhood leukemia risk.

***ACSH assessment*:** The non-quantitative data in this study do not support an environmental link to the childhood leukemia cluster.

26. Nueces County, TX

***NRDC's claim*:** An elevated birth defect rate is associated with exposure and proximity to nearby industrial sites.

***Origin of the claim*:** In 2006, the TDSHS determined that the birth defect rate in Nueces County was 84 percent higher than those of other Texas counties.

***The science*:** A follow-up study of 15 birth defects cases observed between 1996 and 2003 investigated whether these birth defects could be linked to any particular industrial site. While the researchers did find elevations in birth defects as high as 50 percent or greater in areas closest to certain sites, when compared to the incidence in regions farther from the industrial sites, none of the birth defects met all three criteria (high odds ratios [1.50 or greater], statistical significance, and proximity-response [higher odds ratios in mothers living closer to sites]) to meet the TDSHS's standards for a causal relationship between the birth defects and environmental exposure. "My overall bottom line," said senior TDSHS epidemiologist Peter Langlois, "is there is little evidence to say that maternal residential proximity to these sites actually caused the birth defects examined." Langlois also posits that the results of a previous study showing that the birth defect rate in Corpus Christi (within Nueces County) was 17 percent higher than in the rest of Texas can be attributed in part to doctors employing better screening methods for certain defects — particularly minor heart defects — than their counterparts in other parts of Texas.

***ACSH assessment*:** The birth defects are most unlikely to be linked to the proximity of the industrial facilities.

27. San Antonio, Bexar County, TX

***NRDC's claim*:** A liver cancer cluster was caused by polluted groundwater linked to the Kelly Air Force Base.

***Origin of the claim*:** Researchers from Southwest Texas State University found a cluster of liver cancer deaths in Bexar County using statewide cancer mortality data from 1990 through 1997. Local environmental groups allege that the Kelly Air Force Base leached benzene, perchlorate, and TCE into the Bexar County groundwater supply.

***The science*:** Health investigations, however, were not able to establish a causal relationship between the chemicals and the liver cancer fatalities. Though some data on TCE have shown that occupational exposure to TCE may increase the risk of liver cancer, the ATSDR concluded in 2004 that the level of past exposure to the groundwater from the two Kelly Air Force base wells in question is not likely to result in adverse health effects. This conclusion is based on a review of the known toxicology and data on human exposures to the chemicals found in the water. Given the levels of the substances that were detected and the estimated duration of exposures, no adverse health effects are expected to be observed.

***ACSH assessment*:** Past exposure to TCE in the groundwater among Bexar County residents was not high enough to pose a health threat and is not the cause of the elevated liver cancers.

ONGOING OR INCOMPLETE INVESTIGATIONS

28. Middletown-Odessa, DE

***NRDC's claim*:** A cluster of colorectal cancer (CRC) may be linked to an unknown environmental chemical exposure.

***Origin of the claim*:** The DDPH found an elevated incidence of both CRC and esophageal cancer (compared to the total state incidence) among males between 2001 and 2005.

***The science*:** Environmental links were not investigated, and the DDPH is continuing to monitor the area.

***ACSH assessment*:** The NRDC does not specify a chemical exposure, and the cause of the cluster has not yet been investigated.

29. Camp Lejeune, NC

***NRDC's claim*:** Birth defects, childhood cancer, non-Hodgkin's lymphoma, and male breast cancer is associated with exposure to water contaminated with perchloroethylene (PCE), TCE, and benzene.

***Origin of the claim*:** A number of lawsuits filed by former residents of the base who eventually developed cancer and other maladies blamed the marine base's water contamination as the source of their illnesses. The tap water used by United States Marine Corps service members and their families from 1957 to 1987 contained detectable levels of volatile organic compounds (VOCs), such as PCE, TCE, and benzene. NRDC claims that the water contamination from an off-base dry cleaner (PCE), use of trichloroethylene from industrial solvents, and fuel tank leakage of benzene caused a cluster of birth defects, childhood cancer, and male breast cancer. NRDC also bases its claims on media reports that about 60 men who had lived on the base have been diagnosed with male breast cancer.

The science: ATSDR is currently conducting a study on various birth defects, childhood leukemia and non-Hodgkin's lymphoma cases in children born to mothers who lived in Camp Lejeune at any time during their pregnancies. ATSDR is in the process of conducting a health survey that will investigate the incidence of cancer and other diseases, including male breast cancer.

A 1997 ATSDR assessment had found that the VOC levels found in the drinking water were "several thousand times lower than levels of concern in animal studies. They were also hundreds of times lower than levels linked to health effects found in workplace studies" and, therefore, unlikely to have posed a health threat to adults. But ATSDR withdrew this report in April 2009 after acknowledging that the agency overlooked data showing that the water in question contained benzene. Other reports that denied a link between the VOCs and the documented health complications were later determined to have been flawed.

ACSH assessment: Final assessment of the validity of this possible cluster awaits the completion of the ATSDR's ongoing investigation.

30. Clyde, OH

NRDC's claim: A childhood cancer cluster may be linked to an unknown environmental chemical exposure.

Origin of the claim: In 2009, the Ohio Department of Health (ODH) and the Sandusky County Health Department confirmed a cancer cluster in the city of Clyde and Green Creek Township area. The analysis found brain and other central nervous system cancers to be the most common cancer types. NRDC, however, does not provide a particular cause.

The science: State and local health departments continue to investigate what could have caused the elevated number of childhood cancers.

ACSH assessment: The cause of the elevated cancer cases is still being investigated, but no chemical exposures have been implicated.

31. Wellington, OH

NRDC's claim: A multiple sclerosis (MS) cluster can be attributed to a release of volatile organic compounds (VOCs) into the air by a former foundry called LESCO and contamination of groundwater with VOCs by a still operational Forest City Technologies plant, which manufactures automotive parts such as gaskets and seals.

Origin of claim: In 1998, the ODH and Lorain County Health Department identified 25 definite probable cases of MS in Wellington, which consists of approximately 4,200 people. This comes out roughly to a rate of 595 cases per 100,000 people, which is considerably higher than the national average of 58-160 cases per 100,000 people. They also found that Wellington residents were 3.7 times more likely to develop MS than the rest of the country. ATSDR was asked to investigate because of concern about "the possibility of contamination from the Sterling Foundry landfill impacting the adjacent Wellington water reservoir and local soils."

The science: After expanding their MS surveillance to the entirety of Lorain Country, the local and state health departments found that the countywide prevalence of MS was "comparable to that of a similar population in the United States." However, this prevalence study was not intended to determine the cause of the MS cluster. The ATSDR is conducting its own MS case-control study that includes Lorain County as a follow-up to their ongoing MS prevalence study. Multiple studies (including occupational health investigations) have failed to support the association of organic solvents and MS. Genetic studies, on the other hand, have shown that MS occurs more often in family members than in the general population, though the risk for an identical twin is less than 100 percent.

ACSH assessment: The cause of the MS cluster has not yet been determined, and an ATSDR MS investigation is pending.

32. Carbon, Luzerne, and Schuylkill Counties, PA

NRDC's claim: Polycythemia vera (PV) clusters in Carbon, Luzerne, and Schuylkill Counties are linked to multiple "known or suspected carcinogens" associated with a nearby coal-fired power plant and recycling facility.

Origin of the claim: In 2008, ATSDR confirmed a cluster of PV in Schuylkill, Luzerne, and Carbon counties. NRDC reports that some residents blame a nearby coal-fired power plant, in addition to a recycling facility that accepted "thousands of gallons of paint, sludge, waste oils, used solvents, PCBs, cyanide, pesticides, and many other known or suspected carcinogens."

The science: PV is a rare pre-cancerous bone marrow disease that causes the body to make too many red blood cells. The most common consequences of PV are blood clots, heart attacks, and strokes. So far, researchers have been unable to determine what causes PV. ATSDR was charged with confirming the cases associated with the reported cluster of PV, as well as determining whether the cases could be linked to a common cause. Following two rounds of community health screening in northeastern Pennsylvania for the recently discovered JAK2 genetic marker (shown to be present in 95 percent of PV patients), 19 (1.6 percent) of the 1,170 persons tested were found to have this mutation. However, none of these cases could be linked to a common exposure, lifestyle choice, occupation, or ancestry.

ACSH assessment: More research is needed to determine the cause of PV and whether environmental exposures may play a role in the disease. Thus far, no single cause, including an environmental exposure, has been found that could account for the PV cases.

33. Wilkes-Barre, Luzerne County, PA

NRDC's claim: A cluster of 12 Non-Hodgkin's lymphoma and lupus cases in a Wilkes-Barre special education facility, Luzerne Intermediate Unit 18 (LIU 18), is linked to the trichloroethylene (TCE) used to clean two printing presses.

Origin of the claim: A consultant from the Penn State University calculated that the theoretical cancer risk associated with occupational exposure to TCE at LIU 18 for at least 10 years was "10,000 times higher than what the EPA considers acceptable for someone working in the building for at least 10 years." Employees at LIU 18 who worked near the printing presses reported that the TCE often spilled onto the floor and could have infiltrated the ventilation system.

The science: While there are a number of human epidemiological studies that have found an association between high levels of occupational exposure to TCE and non-Hodgkin's lymphoma (a type of cancer of the lymphatic system, which includes the lymph nodes, spleen, and other organs of the immune system), causation has not yet been established. There is also a paucity of occupational data on a TCE link to lupus, a chronic autoimmune disorder that may affect the skin, joints, kidneys, and other organs. The most recent update to the EPA's draft assessment of TCE toxicity does not link TCE to lupus, but there is some evidence to suggest that high levels of exposure to TCE may increase the risk of developing non-Hodgkin's lymphoma.

Conclusion: Given the latest EPA evaluation of TCE, it is possible that the TCE exposure in the 18 LIU workers could have led to a cluster of non-Hodgkin's lymphoma, but not lupus. Governmental inquiry into the cluster is ongoing.

34. Midlothian, TX

NRDC's claim: Air pollution from three cement plants and a steel recycling mill caused a Down syndrome cluster.

Origin of the claim: Residents allege that air pollution from three cement plants and a steel recycling mill caused animal birth defects as well as a Down syndrome cluster between 1991 and 1994.

The science: A governmental health investigation charged with determining the cause of the Down syndrome, along with the allegedly increased incidence of leukemia, birth defects, childhood cancer, and respiratory problems reported by residents is still underway.

***ACSH assessment*:** ACSH will await the results of the currently ongoing investigation and re-assess this alleged cluster.

35. Kenton, DE

***NRDC's claim*:** An elevated cancer rate between 2000 and 2004 is linked to an unknown environmental chemical exposure.

***Origin of the claim*:** The DDPH found a higher rate of many types of cancer in Kenton from 2000 to 2004.

***The science*:** The state investigation did not include research into possible environmental causes of the cluster; the area is still being monitored.

***ACSH assessment*:** The cause(s) of the elevated cancer rate has not yet been determined, and no environmental exposures have been implicated.

36. New Castle County, DE

***NRDC's claim*:** An elevated all-cancer rate in six New Castle County towns is linked to an unknown environmental chemical exposure.

***Origin of the claim*:** The DDPH found an elevated overall cancer rate between 2000 and 2004.

***The science*:** The state health investigations did not look into environmental causes, and the areas are still being monitored.

***ACSH assessment*:** The cause(s) of the elevated cancer rates has not been determined, and an environmental exposure has not been implicated.

Summary

Among NRDC's putative 42 toxic clusters of chemical-induced disease, the current ACSH investigation found that only five had any possible connection to chemical exposure.

Another 13 were inconclusive and those studies are still ongoing. The bulk of the impugned sites — 24 — were based on ideology rather than sound science.

The general rule for risk assessment can be represented by the acronym RITE: **R**isk **I**s equal to **T**oxicity times **E**xposure. In essence, the NRDC assessments of chemical safety have addressed only half of the risk equation — toxicity (or hazard). The NRDC's emphasis on toxicity alone is the hallmark of the precautionary principle; however, in truth, anything can be considered toxic — it all depends on the level of exposure. As we at ACSH have pointed out in our *Dispatch*, most recently on June 20, 2011:

> Everything under the sun, even water, can be considered toxic, or a hazard, to a varying degree. A substance such as cyanide is highly toxic, so even a small exposure creates a risk. Water, on the other hand, is not toxic at common exposures — but too much of it can kill. The precautionary principle doesn't take into account the level

of exposure necessary to present a real risk to health. So when activists call a chemical 'a carcinogen,' they don't take real-world levels of exposure into account.

In reality, the chemical exposures described in the NRDC's report were not at sufficiently high levels, in almost each of their cited instances, to pose an actual health risk.

The Toxic Substances Control Act (TSCA) is the primary law overseeing the safety of chemical products; it grants the U.S. Environmental Protection Agency (EPA) authority to review and regulate chemicals in commerce. TSCA was designed to ensure that products are safe for their intended use. Over time, however, individual state legislatures have created their own chemical regulatory structures, and retailers have pulled products from the shelves — often based on the claims of activists and politicians rather than scientific conclusions. As a result, the regulatory landscape has become a patchwork, a situation that is counterproductive for consumers, the environment, and manufacturers.

Our assessment shows that TSCA has adequately protected U.S. residents from hazardous chemical exposures since its inception. The fact that NRDC chose to include so many "clusters" that were ultimately refuted by rigorous government studies — as well as others whose chemical contamination occurred prior to TSCA's implementation in 1976 — proves that their report was merely one of many agenda-driven attempts to undercut the safety of the products of American industry by blaming manufacturers and their chemicals for the country's maladies.

If Congress is determined to revise TSCA to ensure product safety and consumer confidence, such modernization must be derived from solid science-based principles rather than the "concerns" of activists and agenda-driven politicians and regulators. To ensure confidence in safety regulations, any new law must be based on a strong scientific framework grounded in high-quality data. We at ACSH found that the Natural Resources Defense Council's report fails to provide such data.

1 Bradford Hill, Austin. "The Environment and Disease: Association or Causation?," **Proceedings of the Royal Society of Medicine**, 58 (1965), 295-300.

2 Guzelian PS, et al. **Evidence-based toxicology: a comprehensive framework for causation**. 2005. Hum Exp Toxicol, 24(4): 161-201.

Appendix I

THE BRADFORD HILL CRITERIA

The Bradford Hill criteria, otherwise known as Hill's criteria for causation, are a group of minimal conditions necessary to provide adequate evidence of a causal relationship between an incidence and a consequence, established by the English epidemiologist Sir Austin Bradford Hill (1897–1991) in 1965.

Criteria for Causation

Strength of the association
How large is the effect?

The consistency of the association
Has the same association been observed by others, in different populations, using a different method?

Specificity.
Does altering only the cause alter the effect? (This relates to controlling for confounders.)

Temporal relationship
Does the cause precede the effect?

Biological gradient
Is there a dose response?

Biological plausibility
Does it make sense?

Coherence
Does the evidence fit with what is known regarding the natural history and biology of the outcome?

Experimental evidence
Are there any clinical studies supporting or refuting the association?

Reasoning by analogy
Is the observed association supported by similar associations?

Bradford Hill, Austin. "The Environment and Disease: Association or Causation?," *Proceedings of the Royal Society of Medicine,* 58 (1965), 295-300.

Appendix II

CLUSTERS BY CATEGORY

I. *Valid clusters*

19. Libby, Montana

II. *Possible clusters*

1. Prairie Grove, AR
9. Immokalee, FL (and North Carolina)
16. Midland and Saginaw, MI
20. Bynum, Chatham County, NC

III. *Ongoing or incomplete investigations*

28. Middletown-Odessa, DE
29. Camp Lejeune, NC
30. Clyde, OH
31. Wellington, OH

32. Carbon, Luzerne, and Schuylkill Counties, PA
33. Wilkes-Barre, Luzerne County, PA
34. Midlothian, TX
35. Kenton, DE
36. New Castle County, DE

IV. *Purported clusters lacking supportive evidence*

2. Earlimart, McFarland, and Rosamond, CA
3. Hinkley, CA
4. Kettleman City, CA
5. Montecito, CA
6. Oroville, CA
7. Santa Susana, CA
8. Millsboro, DE
10. Loxahatchee, FL
11. Tallevast, FL
12. Amelia, LA
13. Coteau, LA
14. Mossville, Calcasieu Parish, LA
15. New Orleans, LA
17. White Lake, Muskegon County, MI
18. Herculaneum, MO
21. Marion, OH
22. Marysville, OH
23. Dickson, TN
24. El Paso, TX
25. Houston, TX
26. Nueces County, TX
27. San Antonio, Bexar County, TX

Table I

ACRONYMS

ABC American Beryllium Company
ACSH American Council on Science and Health
ALS Amylotrophic Lateral Sclerosis
ATSDR Agency for Toxic Substances and Disease Registry
BD butadiene
CalEPA California Environmental Protection Agency
CCHD Collier County Health Department
CDC Centers for Disease Control and Prevention
CDHP California Department of Public Health
CDHS California Department of Health Services
CRC colorectal cancer
DDPH Delaware Department of Public Health
DLC dioxin-like chemicals
EMFs electromagnetic fields
EPA Environmental Protection Agency
FLDOH Florida Department of Health
GIS geographical information system
MDEQ Michigan Department of Environmental Quality
MDHSS Missouri Department of Health and Senior Services
MSP Marine Shale Processor
NDCC National Disease Cluster Coalition
NRDC Natural Resources Defense Council
ODH (Ohio Department of Health)
PAHS polycyclic aromatic compounds
PCBs Polychlorinated biphenyls
PCE perchloroethylene
PG&E Pacific Gas & Electric
PV Polycythemia vera
RVHS River Valley High School (Marion, OH)
SSFL Santa Susana Field Laboratory
TCE trichloroethylene
TDSHS Texas Department of State Health Services
TSCA Toxic Substances Control Act
UWF United Farmworkers
VOCs volatile organic compounds

Bibliography

Arkansas Sources

Pell, Sheila. (2010). "Arsenic and old studies." *E — The Environmental Magazine.* [http://www.emagazine.com/archive/5064].

Scott, Janny. (1990. May 9). "State study finds no link of cancer to farm areas." *Los Angeles Times.* [http://articles.latimes.com/1990-05-19/news/mn62_1_childhood-cancer rate].

Wright, A. (2002). "Health department reports low cancer rates, except for testicular, in prelim report to Prairie Grove citizens April 9." Arkansas Department of Health. [http://web.archive.org/web/20020613083905/www.healthyarkansas.com/news/prairiegrove_update_040902.html].

California Sources

California Birth Defects Monitoring Program and California Department of Public Health. (2010). "Birth defects in Kettleman City." [http://www.cdph.ca.gov/Documents/KettlemanCityExternal%20ReportEnglish.pdf].

California Environmental Protection Agency and California Department of Public Health. (2010). "Investigation of Birth Defects and Community Exposures in Kettleman City, CA." [www.calepa.ca.gov/EnvJustice/Documents/2010/KCDocs/ReportFinal/Final-Report.pdf].

Coye, M. & L.R. Goldman. (1991). "Summary of environmental data: McFarland childhood cancer cluster investigation." Environmental Epidemiology and Toxicology Program, California Department of Health Services. [http://www.ehib.org/papers/10_Coye_1991_McFarland.pdf].

Harnly, Martha. (2002). "Questions and answers about childhood cancer in Rosamond." California Department of Health Services, Environmental Health Investigations Branch. [http://www.ehib.org/papers/rosamond_cancer.pdf.]

Kreutzer, R. (1993). "Investigation of the Earlimart childhood cancer cluster, California Department of Health Services." Environmental Health Investigations Branch. [http://www.ehib.org/papers/19_Kreutzer_1993_Invest_F24.pdf].

Kreutzer, R., R. Schlag, E. Glazer, G. Micarelli, E. Blake, & L. R. Goldman. (1990). "Investigation of the Montecito leukemia and lymphoma cluster, final report." Environmental Epidemiology and Toxicology Program, California Department of Health Services. [http://www.ehib.org/papers/29_Kreutzer_1990_Invest_059.pdf.]

Morgan, John W. (2011). "Preliminary assessment of cancer occurrence in the Hinkley census tract, 1996-2008." California Cancer Registry.[http://library.constantcontact.com/download/get/file/1102141380567S8/20110323_Hinkley_3_2011.pdf].

Morgenstern, Hal, Beebe-Dimmer, Jennifer, & Yu, Sunkyung. (2007). "Cancer Incidence in the Community Surrounding the Rocketdyne Facility in Southern California." Final Report to Eastern Research Group, Lexington, MA. [http://www.ph.ucla.edu/erg/final_epi_report.pdf].

Reynolds, P., Smith D.F., Satariano E., Nelson D.O., Goldman L.R., & Neutra, R.R. (1996). "The four county study of childhood cancer: Clusters in context." *Statistics in Medicine.* 15(7-9): 683-697.

Santa Susana Field Laboratory Advisory Panel. (2006). "Report of the Santa Susana Field Laboratory Advisory Panel." [www.ssflpanel.org/files/SSFLPanelReport.pdf].

Smith, D., Hoshiko, S., Naik, K., Fan, C., Fitzsimmons K., & Rojas-Cheatham, T. (2009) "Pancreatic cancer follow-up investigation report." California Department of Public Health, Environmental Health Investigations Branch. [www.ehib.org/projects/OrovillePancreaticCancerInvestigation.pdf].

Stimola, Aubrey. (2004). "Chromium and the Erin Brockovich effect." American Council on Science and Health. [http://www.acsh.org/factsfears/newsID.326/news_detail.asp].

Delaware Sources

Delaware Division of Public Health. (2010) "Cancer cluster investigations in Delaware." [http://dhss.delaware.gov/dph/dpc/clusterinvestigations.html].

Delaware Health and Social Services Division of Public Health. (2008). "Average annual age-adjusted cancer incidence rates, 2000-2004, at the Delaware sub-county level." [http://www.dhss.delaware.gov/dhss/dph/dpc/files/subcounty_cancerincidencereport.txt].

Florida Sources

Agency for Toxic Substances and Disease Registry. (2008). Public Health Assessment for Former American Beryllium Co. [www.atsdr.cdc.gov/HAC/pha/FormerAmericanBeryllium/American_Beryllium_Company%20PHA%209-30-2008.pdf].

Calvert, Geoffrey M., Alarcon, W. A., Chelminski, A., Crowley, M. S., Barrett, R., Correa, A., Higgins, S., Leon, H. L., Correia, J., Becker, A., Allen, R. H., & Evans, E. (2007). "Case report: Three farmworkers who gave birth to infants with birth defects closely grouped in time and place — Florida and North Carolina, 2004–2005." *Environmental Health Perspectives, 115* (5). [http://www.ncbi.nlm.nih.gov/pmc/articles/PMC1867963/].

Chelminski, Ann N. & Higgins, Sheila. (2006). "Assessment of maternal occupational pesticide exposures during pregnancy and three children with birth defects: North Carolina, 2004." Occupational and Environmental Epidemiology Branch Division of Public Health North Carolina Department of Health and Human Services. [http://www.epi.state.nc.us/epi/oii/Agmartreleasereport.pdf].

Florida Department of Health, Division of Disease Control, Bureau of Epidemiology. (2010). "Acreage community case-control data analysis."[http://www.pbchd.com/pdfs/acreage/acreage_community_casecontrol_data_analysis_report-9_13_2010.pdf].

Layden, L, & Zeitlin, J. (2005, October 13). "Health officials: Pesticides not likely at fault for birth defects." *Naples Daily News*. [http://www.naplesnews.com/news/2005/oct/13/health_officials_pesticides_not_lkely_fault_birth/].

Malek, M. (2009, July 27). "Acreage residents worry Pratt & Whitney contaminants played role in brain tumors." *Palm Beach Post*.[www.palmbeachpost.com/health/content/local_news/epaper/2009/07/27/0723pratt.html].

McGeehin, M.A. Centers for Disease Control and Prevention. (2010) Acreage Investigation Letter to Lisa Conti, Division of Environmental Health, Florida Department of Health [www.pbchd.com/spotlight/acreage/cancer_cluster.html].

Louisiana Sources

Agency for Toxic Substances and Disease Registry. (1994). "Petitioned Public Health Assessment: Marine Shale Processors, Inc." [http://www.atsdr.cdc.gov/HAC/pha/PHA.asp?docid=725&pg=0].

Agency for Toxic Substances and Disease Registry. (2006). "ATSDR study finds dioxin levels in Calcasieu Parish residents similar to national levels."[http://www.atsdr.cdc.gov/news/displaynews.asp?PRid=1847].

Agency for Toxic Substances and Disease Registry. (2006)."Health Consultation: Follow-up exposure investigation, Calcasieu Estuary (a.k.a Mossville)." [http://www.atsdr.cdc.gov/hac/pha/CalcasieuEstuary/CalcasieuEstuaryHC031306.pdf].

Agency for Toxic Substances and Disease Registry. (2007). "Health consultation: Assessment of cancer incidence from the Louisiana Tumor Registry from 1988 –2004, Calcasieu Parish, Louisiana." [http://www.atsdr.cdc.gov/hac/pha/CalcasieuCancer/CalcasieuCancerHC92707.pdf].

Agency for Toxic Substances and Disease Registry. (2009). "Health assessment: Agriculture street landfill." [http://www.atsdr.cdc.gov/HAC/pha/pha.asp?docid=707&pg=1].

Kelly, K. E. (1995). "Health effects of hazardous waste incineration ... more of the rest of the story: An updated review of the scientific basis of alleged adverse health effects of hazardous waste incineration." Presented at the International Incineration Conference, May 1995. [http://www.deltatoxicology.com/pdf/5cases.pdf].

Louisiana Department of Health and Hospitals. (2000). "2000 Louisiana health report card: Health assessment programs." [http://www.dhh.louisiana.gov/offices/miscdocs/docs-275/recordsstatistics/statistics/docs/reportcards/2000/III.%20Health%20Assessment%20Programs.doc].

Martin, D. S. (2010. February 26). "Toxic towns: People of Mossville 'are like an experiment.'" CNN Health. [http://www.cnn.com/2010/HEALTH/02/26/toxic.town.mossville.epa/index.html].

Mossville Environmental Action Now, Incorporated, Lake Charles Area Concerned Citizens, Communities for a Better Environment - NORAN Project, Earthjustice Legal Defense Fund (Louisiana Office). (2000). "Breathing poison: The toxic costs of industries in Calcasieu Parish, Louisiana," Section 2.2. [http://www.mapcruzin.com/mossville/reportondioxin.htm].

Michigan Sources

Agency for Toxic Substances and Disease Registry. (2008). "ATSDR studies on chemical releases in the Great Lakes region." [www.atsdr.cdc.gov/grtlakes/pdfs/2008/final/Chapter5_GreatLakesAOCFinal.pdf].

Agency for Toxic Substances and Disease Registry. (2008). "Health consultation: Evaluation of Saginaw River dioxin exposures and health risks." [http://www.michigan.gov/documents/mdch/Saginaw_River_HC_Final_12908_1_223827_7.pdf].

Dai, D. &. Oyana, T.J. (2008). "Spatial variations in the incidence of breast cancer and potential risks associated with soil dioxin contamination in Midland, Saginaw, and Bay Counties, Michigan, USA." *Environmental Health*, 7: 49. [www.ncbi.nlm.nih.gov/pubmed/18939976].

White Lake Beacon. (2011, April 18). "White Lake area community leaders rebut cancer cluster designation." [http://www.whitelakebeacon.com/news/24089-no-cancer-cluster].

Missouri Sources

Missouri Department of Health and Senior Services Office of Epidemiology. (2007). "Determining baseline prevalence for provider-diagnosed multiple sclerosis (MS) and amyotrophic lateral sclerosis (ALS) in Herculaneum and Jefferson County, Missouri." [http://www.dhss.mo.gov/living/environment/hazsubstancesites/pdf/FinalReport_H erculaneum.pdf].

Montana Sources

Agency for Toxic Substances and Disease Registry. (2002). "Health consultation: Mortality from asbestosis in Libby, Montana, 1979-1998." http://www.atsdr.cdc.gov/hac/pha/pha.asp?docid=1225&pg=0

North Carolina Sources

Agency for Toxic Substances and Disease. (2009). "Study on birth defects and childhood cancers," [www.atsdr.cdc.gov/sites/lejeune/update.html].

Newsobserver.com. (2011, March 30). "Bynum cancer cluster resolved years ago, Chatham health director says. [http://www.newsobserver.com/2011/03/30/1092490/bynum-cancer-clusterresolved.html.]

Osbourne, J. S. II, Shy, C.M., & Kaplan. B. H. (1990). "Epidemiologic analysis of reported cancer cluster in a small rural population." *American Journal of Epidemiology, 132*, Suppl., No. 1. [www. ncbi.nlm.nih.gov/pubmed/2356841].

Ohio Sources

Agency for Toxic Substances and Disease Registry. (2005). Health Consultation: Village of Wellington [www.atsdr.cdc.gov/hac/pha/VillageofWellington031705-OH/Villageof-Wellington031705-OH.pdf].

Associated Press. (2001, July 28). "A Town Divided: How a cancer threat is tearing a community apart." [http://www.ohiocitizen.org/campaigns/marion/inconclusive.htm].

Clark, B.R., Ferketich, A.K., Fisher, J.L., Ruymann, F.B., Harris, R.E., &Wilkins, J.R. 3rd. (2007). "Evidence of population mixing based on the geographical distribution of childhood leukemia in Ohio." *Pediatric Blood Cancer, 49* (6), 797-802.

Ohio Department of Health, Sandusky County Health Department, & Ohio Environmental Protection Agency. (2009). "Childhood cancer among residents of eastern Sandusky County." Progress Report. [http://www.epa.state.oh.us/LinkClick.aspx?fileticket=UyvHDgC4wG8%3D&tabid=3253].

US Army Corp of Engineers. (2010). "Marion Engineer Depot — Marion, Ohio." [http://www.lrl.usace.army.mil/poi/default.asp?mycategory=255].

Pennsylvania Sources

Agency for Toxic Substances and Disease Registry. (2010). Community health screening report for JAK 2 (V617F) mutation: Luzerne, Schuylkill, and Carbon Counties, Pennsylvania. [www.atsdr.cdc.gov/HAC/pha/CommunityHealthScreeninginPA2010/CommunityHealthScreeningReport.pdf].

Morris, J. (2005, January 11). "Widely used chemical might have cancer link." *Boston Globe.* [http://forums.lymphoma.com/archive/index.php/t-14669.html].National Academy of Science. (2006). *Assessing the human health risks of trichloroethylene: key scientific issues.* The National Academies Press. http://www.nap.edu/catalog.php?record_id=11707.

National Research Council & National Academy of Science. (2006). "Report in brief: assessing the human health risks of trichloroethylene: Key scientific issues." [http://dels.nas.edu/resources/static-assets/materials-based-on-reports/reports-in- brief/trichloroethylene_brief_final.pdf].

Preuss, P. W. (2010). "U.S. EPA's draft toxicological review of trichloroethylene (TCE): Presentation for the Science Advisory Board." National Center for Environmental Assessment. [http://yosemite.epa.gov/sab/sabproduct.nsf/C4E2D10DE301736C8525772900760E71/$File/TCE_Preuss+Presentation+to+TCE+Panel+for+My+10-12+2010+Meeting.pdf].

Tennessee Sources

Bove, F.J., Fulcomer, M.C., Klotz, J.B., Esmart, J., Dufficy, E.M., Savrin, J.E. (1995)."Public drinking water contamination and birth outcomes." *American Journal of Epidemiology, 141* (9), 850-862.

Conner, K. (2000, September 27). "Study points to birth defects cluster." *The Dickson Herald.* [http://www.epa.gov/region4/foia/readingroom/dickson_county/documents/article_sept_2000_29.pdf].

Letter from Dana C. Crawford, EIS Officer to Director, Divison of Training, EPO. Re: Epi-Aid trip report: Possible cluster of orofacial clefts. (2001). [http://www.epa.gov/region4/foia/readingroom/dickson_county/documents/epi_aid_trip_report_jan_2000_2.pdf].

Tennessee Government, Department of Environment & Conservation. Division of Solid and Hazardous Waste Management. [http://www.tn.gov/environment/swm/].

National Academy of Science. (2006). *Assessing the human health risks of trichloroethylene: Key scientific issues.* The National Academies Press. [http://www.nap.edu/catalog.php?record_id=11707].

National Research Council & National Academy of Science. (2006). "Report in brief: assessing the human health risks of trichloroethylene: Key scientific issues." [http://dels.nas.edu/resources/static-assets/materials-based-on-reports/reports-in-brief/trichloroethylene_brief_final.pdf].

Preuss, P.W. (2010). "U.S. EPA's draft toxicological review of trichloroethylene (TCE): presentation for the Science Advisory Board." National Center for Environmental Assessment. [http://yosemite.epa.gov/sab/sabproduct.nsf/C4E2D10DE301736C8525772900760E71/$File/TCE_Preuss+Presentation+to+TCE+Panel+for+My+10-12+2010+Meeting.pdf].

Hannigan, J.H., Bowen, S.E. (2010). "Reproductive toxicology and teratology of abused toluene." *Systems Biology in Reproductive Medicine, 56* (2), 184-200. http://www.ncbi.nlm.nih.gov/pubmed/20377315

Texas Sources

Agency for Toxic Substances and Disease Registry. (2001). "Health consultation: Past exposure to drinking water from on-base wells 313 and 314 Kelly Air Force Base." [http://www.atsdr.cdc.gov/HAC/pha/KellyAFBPC102804/KellyAFB_Onbase_Drinking_Water.pdf].

Agency for Toxic Substances and Disease Registry & Texas Department of State Health Services. (2007)."Executive summary: Health consultation, Midlothian area air quality, Part 1: volatile organic compounds & metals, Midlothian, Ellis County, Texas." [http://www. dshs.state.tx.us/epitox/midlothian/executive_summary.pdf.]

Agency for Toxic Substances and Disease Registry. (2007). "Health consultation, Midlothian area air quality, Part 1: Volatile organic compounds & metals, Midlothian, Ellis County, Texas," [http://www.dshs.state.tx.us/epitox/midlothian/midlothian_airquality_consult_121107.pdf].

Agency for Toxic Substances and Disease Registry. (2011). "Health Consultation: Exposure Investigation, Biological and Environmental Monitoring for Exposure to Benzene and Related Petroleum Chemicals, Corpus Christi, Texas." [http://www.atsdr.cdc.gov/HAC/pha/CorpusEI/CorpusTXHCEI12052011.pdf].

Langlois, P. Texas Department of State Health Services. (2008) "A case-control study of the association between birth defects elevated in Nueces County and sites of concern to citizens for environmental justice" [http://www.caller2.com/2008/pdf/0125defectsnc2.pdf].

Linder, S. H., Marko, D., & Sexton, K. (2008). "Cumulative cancer risk from air pollution in Houston: Disparities in risk burden and social disadvantage." *Environmental Science & Technology, 42* (12).

Ross, G. "Houston can breath easy over UT study." (2007, February 24). *The Houston Chronicle.* [http://www.chron.com/disp/story.mpl/editorial/outlook/4579395.html].

Texas Department of State Health Services. (2011). "El Paso multiple sclerosis study." [www.dshs.state.tx.us/epitox/elpasostudy.shtm].

Zhan, F. B. (2002). "Are deaths from liver cancer, kidney cancer, and leukemia clustered in San Antonio?" *Texas Medicine, 98* (10). [http://uweb.txstate.edu/~fz01/Reprints/Zhan_2002_TexMeD.pdf].

Miscellaneous Sources

Bradford Hill, A. (1965). "The environment and disease: Association or causation?" **Proceedings of the Royal Society of Medicine, 58**, 295-300.

Guzelian P.S., **et al**. (2005). **Evidence-based toxicology: a comprehensive framework for causation**. 2005. **Human & Experimental Toxicoly, 24** (4), 161-201.

Navarro, K., Janssen, S., Nordbrock, T., & Solomon, G. (2011). "Health alert: Disease clusters spotlight the need to protect people from toxic chemicals." [http://www.nrdc.org/health/diseaseclusters/files/diseaseclusters_issuepaper.pdf].

BOARD OF SCIENTIFIC AND POLICY ADVISORS *(continued)*

BOARD OF SCIENTIFIC AND POLICY ADVISORS *(continued)*

BOARD OF SCIENTIFIC AND POLICY ADVISORS *(continued)*

Made in the USA
Monee, IL
04 November 2026

67386124R10036